On thinking with —
scientists, Sciences, and Isabelle Stengers

On thinking with —
scientists, Sciences, and Isabelle Stengers

text by Jeremy Fernando
with illustrations by Maureen Burdock
and a layout by Mariane Klettenhofer

This paperback edition first published
in 2015 by Delere Press LLP

Images © Maureen Burdock
Text © Jeremy Fernando
Layout © Mariane Klettenhofer

First published in 2015 by
Delere Press LLP
Block 370G Alexandra Road
#09-09 Singapore 159960
www.delerepress.com
Delere Press Reg No. T11LL1061K

all rights reserved

ISBN 978-981-09-8990-3

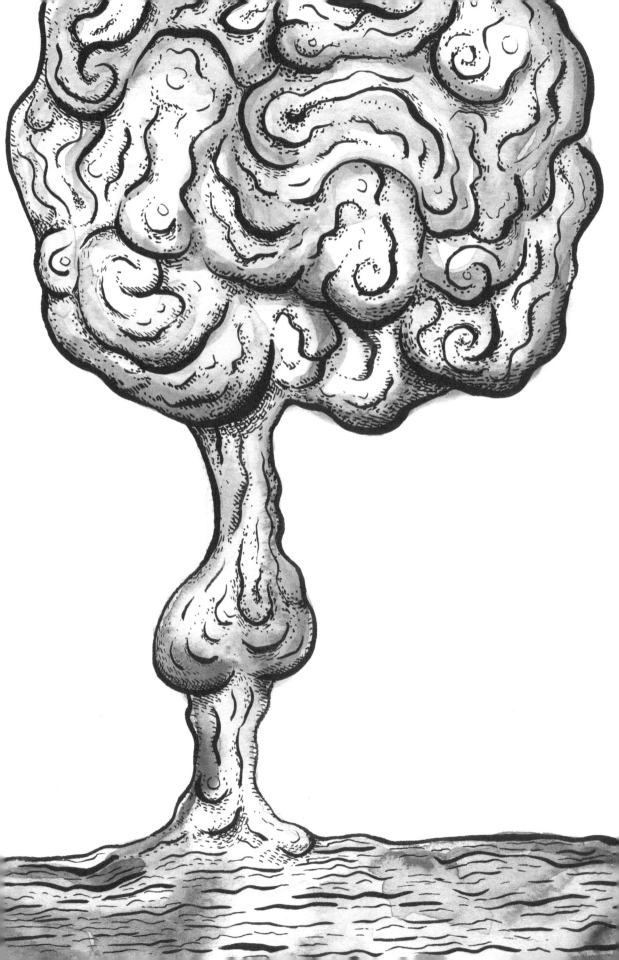

... and out of this speech from the body which I lose again the book will be born, torn to shreds by these marks and commas, but not dead, but a tomb...

Hélène Cixous

[1] This talk took place at noon on 1 June, 2015, at the Centre for Science and Innovation Studies, UC Davis, as part of the Sawyer Seminar series — barely 24 hours after I arrived from Singapore, which is 15 hours ahead in time.

I would like to thank Marisol de la Cadena & Mario Biagoli for their warm hospitality; and Andrew Ventimiglia for so generously hosting the event.

[2] All quotations in orange are taken from a transcript of Isabelle Stengers' talk entitled 'Cosmopolitics — Learning to Think with Sciences, Peoples and Natures' which was delivered at Halifax University in March 2012.

I would like to begin, to open, by saying, declaring — confessing perhaps — that I am about to speak about, have already started speaking of, things, of people, that I know very little of. For, there is indeed very little about Science that I know, less even of scientists, and perhaps even less about Isabelle Stengers.

So perhaps, an opening that asks for, that requests, your indulgence.

But perhaps also, an excuse that comes a little too late. After all, I have already begun, at least in terms of words, scribbles, marks. Not that we can quite know when we ever actually begin.

After all, I might have already started this the moment I agreed to come, to be here — the moment I responded to Mario Biagioli's very kind, deeply generous, invitation to speak on something; a something that neither of us had quite known at the point of agreement, at the moment of the touch of a handshake, what it was, what it might be. Or, if it might even amount to anything.

Perhaps then, I should have begun by asking to begin again. Not that one can ever quite go back in time, at least not in any known way; but that we might open ourselves to the possibility of having multiple tracks in time, keeping in mind that I am currently in two time zones at least;[1] perhaps then having times that open us to — in Isabelle Stengers' words — a weaving of regenerative, slightly transgressive imaginations.[2]

So, perhaps as a third beginning, one that attempts to interweave my attempt to speak as an amateur with the double, triple, multiple times that I am reading Stengers' text in — without quite bringing them together, without flattening the singularity of love (*amore*) and the multiplicities of engagements across particular times into one — perhaps you will allow me to open our ears to the tune that was playing in my head, for reasons that still escape me, the very first time, and each and every time, I read, listen to, Stengers' speech, 'Cosmopolitics — Learning to Think with Sciences, Peoples and Natures', which I am trying to respond to, think with.

And in her spirit, perhaps the question is not so much what the tune signifies, what Tracy Bonham might mean, but its relevance, its effects. After all, there is a possibility that tunes, like ideas, have an efficacy of their own, to poison or to activate, to close or to open possibilities.

A song which opens the question — or at least, a possible question — for us:

WHY TWICE? Why the invocation of *mother* twice over; or, even — why call out to two mothers?

Perhaps here, we might consider the notion that we always already have two mothers: the one from whom, through whom, we arrive, and the one that we call the maternal; the woman, and the mother. Not necessarily separate, but perhaps never quite the same. And since these thoughts are coming through one who calls, names himself — answers to the call of being — South East Asian, perhaps we might consider the possibility that they are *same same but different*.

Where mother is the one who watches over you, that cares for you — that pays attention to you.

Which is precisely what Isabelle Stengers says Gaia is not.

For her, Gaia is the name of a very ancient divinity, a Greek divinity much older than the anthropomorphic Gods and Goddesses of the Greek cities. It may be that

When you sent me off to see the world, were you scared that I might get hurt? Would I try a little tobacco, would I keep on hiking up my skirt?

Tracy Bonham,
Mother mother

she was a figure of the mother, but then not of a nice, loving mother, rather of an awesome one, who should not be offended, also of a rather indifferent one, with no particular interest in the fate of her offspring … Gaia is this figure of the many figured Earth which demands neither love, nor protection but the kind of attention to be paid to a prickly powerful being.

And more than that: for — attending to Stengers' teaching here — to name is a pragmatic operation, the truth of which is in its effects.

Thus, Gaia is an unknown mother, a potentially unknowable mother; yet one that we have to pay attention to, attend to, attempt to respond to. A mother that is a trace, an echo, perhaps always already the second mother, as it were. And also, a name which names nothing except the fact that it is naming.

But perhaps before we attempt to unravel the mysteries of Gaia — keeping in mind that true mysteries make us tremble — we should turn our attention to the situation of her talk, her speech; particularly since specificity is crucial to Stengers.

So, even as we might be reading her — or, reading a reading of her, as it were — we should try not to forget that Stengers was attempting to address an audience composed of those familiar with Science, with the Sciences if you prefer. Thus, it is not so much the content of her speech that is important (which is not to say that what she says isn't crucial, wasn't illuminating — far from it) but that we should pay attention to the effects of it, to the manner in which she attempts to affect those who had listened, were presumably listening. Where, what she was attempting to do was to

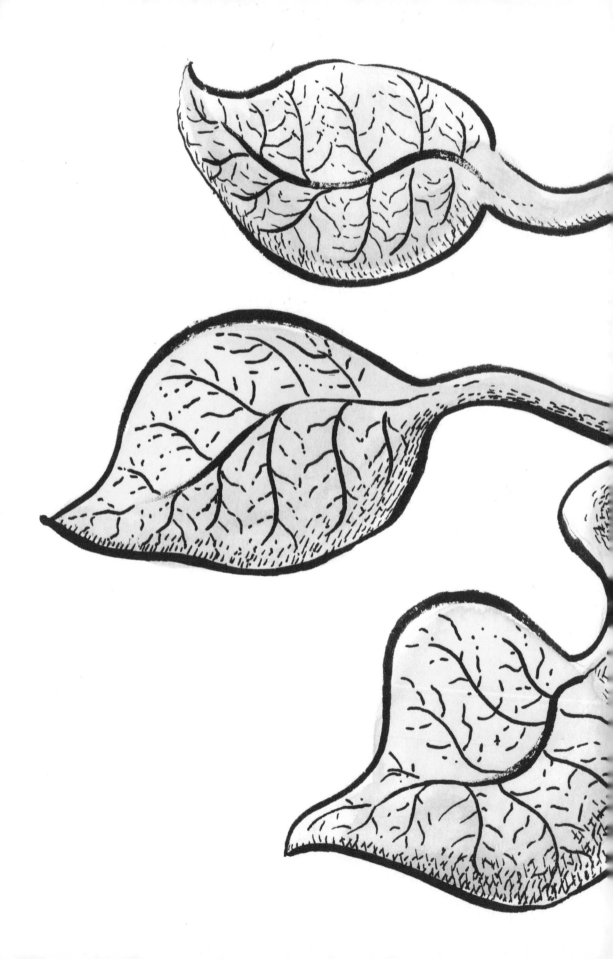

lead her listeners down a particular path. And here, we should open the dossier of pedagogy, particularly the register that the teacher—the *pedagogue*—only can guide, lead (*agogos*; leader) the ones (*paida*; boy) being taught. Where, teaching is not a direct transference of information, or even knowledge, but a leading by example. Where the habits of the teacher—and by extension the teacher's body (*habitus*)—is the very site of the teaching. Which is why Martin Heidegger teaches us that, "the real teacher, in fact, lets nothing else be learned than—learning. His conduct, therefore, often produces the impression that we properly learn nothing from him, if by 'learning' we now suddenly understand merely the procurement of useful information. The teacher is ahead of his apprentices in this alone, that he has still far more to learn that they—he has to learn to let them learn. The teacher must be capable of being more teachable than the apprentices. The teacher is far less assured of his ground than those who learn are of theirs."3 Thus, the teacher and the student are in a relationality, where they are open to the possibility of learning—and where this learning takes place is on, and in, their bodies.

Where what Stengers is attempting to do is: open a relation with her audience; recalling her notion that the making of relation is not the recognition that we are related, it is an achievement, which implies the risk of failure, the hesitation between peace and war. A relationality that doesn't assume its own relation; one in which the one who is leading is not only willing to be lost, but to be led by the very ones she is leading. Where being led astray is not just a possibility, but will always already remain a potentiality. Not because of

3 Martin Heidegger. *What is Called Thinking?*, translated by J. Glenn Gray. New York: Perennial, 15.

Her too Martin—always her as well.

4 Jacques Derrida. *Demeure: Fiction and Testimony*, translated by Elizabeth Rottenberg. Stanford: Meridian, 2000: 27.

Reading here being understood as the relation to an other that occurs prior to any semantic or formal identification, and therefore prior to any attempt at assimilating what is being read to the one who reads. As neither an act nor a rule-governed operation reading, thus, needs to be thought as an event of an encounter with an other — and more precisely an other which is not the other as identified by the reader, but heterogeneous in relation to any identifying determination. Thus, a pre-relational relationality where what the reader encounters may only be encountered before any phenomenon; hence a non-phenomenal event or even the event of the undoing of all phenomenality.
Jeremy Fernando, *Reading Blindly*

misdirection — for, if there is no *telos*, no known goal, one cannot know if one ever reaches a destination, gets there — but because this uncertainty is written into every journey, every path, every *hodos*.

The same unknowability that haunts every method, every *meta hodos*.

In other words, a relationality that is authored, a relationality that cannot exist before being written into being — which also means that this relationality is one that has no necessary legitimacy, *grund*, quite possibly no possibility of verification. A writing that is a reading of — **IS READING** — the possibility of relationality between the two or more things, beings, brought into relation with each other as the relation is being written.

Which is why Stengers calls for witnessing, calls for — to borrow her terms — experts, diplomats, and victims, to testify to the relationship between themselves.

And here — perhaps at the risk of being unfaithful to Stengers' project of being constructivist, pragmatist, speculative, of responding with infidelity to her spirit even — we should not close ourselves to the teachings of Jacques Derrida on testimony, on witnessing; in particular, the notion that fictionality is both the condition and the very limit of testimony; that without the possibility of fiction, there would not be the possibility of testifying to one's experience. As Derrida testifies, "testimony always goes hand in hand with at least the *possibility* of fiction, perjury, lie. Were this possibility to be eliminated, no testimony would be possible any longer; it could not have the meaning of testimony."[4] However, "the essence of testimony cannot necessarily be reduced to narration, that is, to descriptive, informative relations, to knowledge, or to narrative; it is first a present act."5 And, since it has to be a "present act," this suggests that in order to testify to an experience, one has to have first lived through it; thus, all testimony occurs through memory. However, since one has no control over, nor access to, forgetting — it happens to one — one can never be certain not only if one has forgotten anything in the testimony by way of omission, one can also never quite know if every act of memory entails, brings with it, forgetting. Which might also be why there can be unknowingly false — or even true — testimonies. Or, perhaps more aptly, that one can never be certain, or even aware, of the truth or falsity of one's testimony. Moreover, since there is no, or at least no known, limit to the number of testimonies that one can offer regarding a particular event, this suggests that it is precisely forgetting itself that allows this to happen: where, it is the impossibility of verification that allows for each and every testimony.

Which suggests that the risk of all testimony, of testifying, is that one is not responding, reading, but — instead — writing the object of testimony into being; making it speak for one, speak only because one is speaking through it, say what one wants it to say.

PROSOPOPOEIA. Perhaps then, all one can know is that one is testifying. Perhaps all one can ever do is **TO NAME** one's testimony as testimony. Keeping in mind that to name is always already to prepare for the absence of the one who is being named; to prepare for the moment in which all that can be uttered is her, his, its, name — for the moment of its, his, her, death.

5 *Ibid*: 38.

The form of the name — a place of solitary confinement — eats the body and holds it upright.

Jacques Derrida, *Glas*

For this is the way in which religions are wont to die out: under the stern, intelligent eyes of an orthodox dogmatism, the mythical premises of a religion are systematized as a sum total of historical events; one begins apprehensively to defend the credibility of the myths, while at the same time one opposes any continuation of their natural vitality and growth; the feeling for myth perishes, and its place is taken by the claim of religion to historical foundations.

Friedrich Nietzsche, *The Birth of Tragedy*

In other words, to call one's testimony a testimony, to name it as such, is to always also prepare for the moment when it testifies to nothing. A moment perhaps when victims, diplomats, experts, are testifying to nothing except for the fact that they are testifying — that even as they are named testifiers, that even as they testify, their testes have been cut off.

However, even as we tune our ears to Derrida, we should also not forget, not wash our hands of, Stengers' statement that she is not doing a critical deconstruction. For, even if factually justified, deconstruction fails from the pragmatic, speculative point of view, from the point of view of its effects, leaving us with a more desolate, empty world. Thus, I have to acknowledge, bear responsibility — a notion clearly dear to Stengers — to the fact, or at least the possibility, that I am misreading her. Which is not to say — if one is to take seriously the notion that reading is an openness to the possibility of an other — that one can necessarily distinguish, with any certainty, a legitimate from an illegitimate reading. But perhaps, this is the very site of responsibility itself: the acknowledgment that all readings are one's readings, readings that one has written — are always potentially illegitimate readings.

Readings that cannot count on any authority, **CANNOT COUNT ON DADDY** for legitimacy. Readings that can only know who mommy is.

But perhaps, this is precisely what saves the readings, the writings of these readings, relationality itself, from certainty, from being set in stone — the very things that Stengers is warning us about. For in her words:

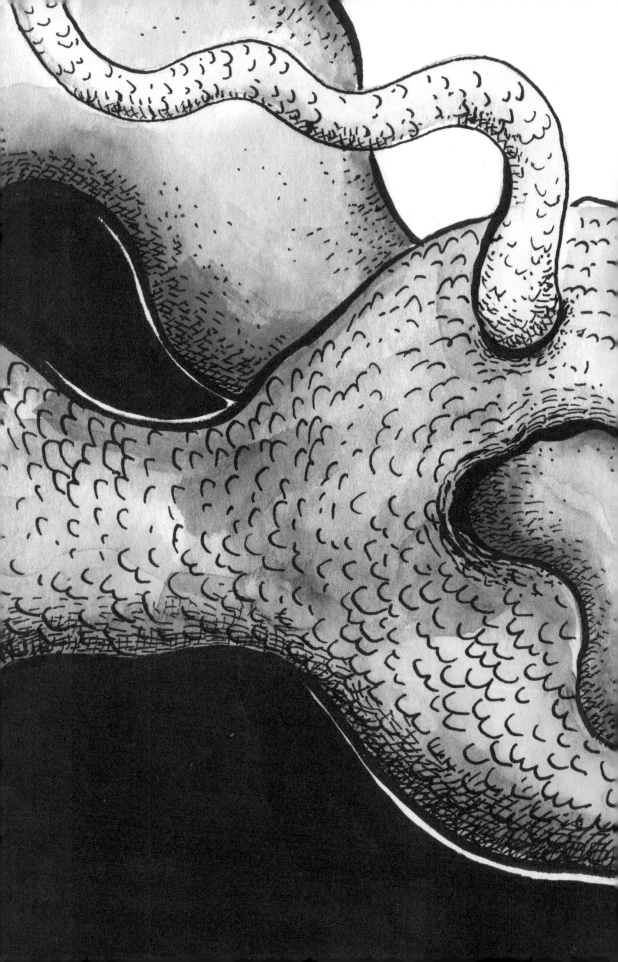

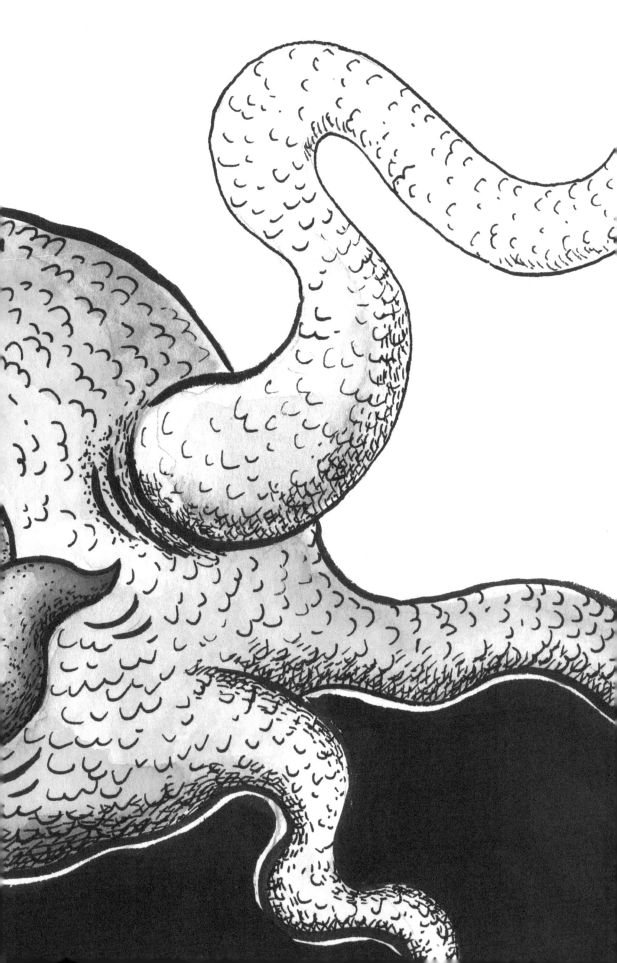

I vitally need
a dream,
such a story
which never happened ...

A dream of vitality, energy, movement; a dream about possibilities. A dream in which the effects are, a positive, radical, plurality of sciences, each particular scientific practice answering the challenge of relevance associated with its field. And in her vital dream, one can hear the whisperings of Nietzsche, in particular his gay scientist: the joyful tester who continues testing, testing the very test itself—never settling on an answer, never allowing an answer to settle. For, what maintains movement, life, vitality, is precisely uncertainty: the possibilities that fictionality opens to another tale, story.

In speaking about religions—a register that should not be foreign to our dossier of missing paternity, of the missing father—Nietzsche teaches us that the moment religion shifts from a movement—from constantly changing, morphing, becoming—into stagnancy, being, dogma, settles into a structure, where all "vitality and growth" are drained from it, it becomes lifeless, dead. And, it is precisely the attempt to fully understand the religion—move it from myths which are dynamic, ever-changing, constantly re-told, altered, alive, to a set story, hi-story, linear, predictable, re-traceable, uncontaminated by variation—that murders it, that concretises it into mere orthodoxy.[6]

And it is this concretisation that Stengers detects—calls out—in the general, unilateral authority of science, conquering the world, defining what really

[6] Friedrich Nietzsche. *The Birth of Tragedy*, translated by Walter Kaufman. New York: Vintage Books, 1967: 75.

Dogma:
Latin for 'philosophical tenet'; from the Greek *dogma* **(genitive** *dogmatos***), 'opinion, tenet' literally 'that which one thinks is true'; from** *dokein* **'to seem good, think'**

Orthodoxy:
from the Late Latin *orthodoxus***; from the Late Greek** *orthodoxos* **'right opinion'; from orthos 'right, true, straight' +** *doxa* **'opinion, praise'; from** *dokein*

matters and what are illusory beliefs only, blessing the destruction of innumerable other ways of relating, knowing, feeling, and interpreting.

The general, who wants to be the leader, who leads his armies, flattening all differences, all otherness, all other possibilities. Who wants to be the source of all knowing, all knowledge — daddy — the origin, *auctor*.

Which is not to say that Stengers' call for plurality, for possibilities, is one that is devoid of influence, of power. For, effects are crucial to her. But here, one should be attentive to Stengers, and bear in mind that it is power that she is speaking of, and not authority. Here, one must also try to not forget that whilst power can be contested, challenged, authority is mystical, divine, outside the realm of human consciousness — it is of the order of the sovereign. One either has authority or one doesn't. And even if we take into account the fact that authority has to be granted by another — for, the moment a figure of authority uses force, violence, power, her authority dissolves, vanishes — we cannot ever know why authority is granted to some and not others. And, when we attempt to offer a reason for figures of authority, we inevitably turn to notions — almost as a last resort — such as charisma. However, we should bear in mind that *kharisma* is a divine gift — it not only comes from elsewhere, but is always also beyond reason, beyond us. Thus, with an origin whose origins remain veiled from us. Or, more than that: an origin whose origins might very well have been authored by us.[7]

And, since the dossier of gift has been opened, if we pay attention, it is not too difficult to hear echoes

[7] I have explored the illusory origins of authority and the appeal of leaders at greater length in a piece entitled 'An essay on what it is to be a leader' in *Vice UK*, April 2013.

of *datum*, of data, resounding: keeping in mind that a *datum* is an unrequitable gift; thus, always also, unequatable, uncalculable, unexchangeable. This suggests that data would involve not just the movement of thought, knowledge, but always also brings with it notions of a sharing that is beyond calculability: therefore — borrowing more contemporary parlance; and here, one should remain attentive to *time*, and how the contemporaneous is not linear but a rupture in linearity itself — data and sharing have always been in relation with each other, data has always already been open source. Which also means that data — sharing, transference — always entail an openness to the possibility of another; along with the potentiality for disruption, infection, viruses, distortion. Thus, even as the data is shared, what exactly is shared remains unknown until the moment it is shared — a repetition, a replication, that is same but which might not quite be the same, even if it is recognised, seen, as such. Perhaps then, even as we posit that data is shared, *what exactly is data* might remain veiled from us. For, we should not forget that a *datum* is a gift from a master to her slave, a gift of freedom, always an unequal gift, a gift that never allows the one receiving to forget not just from whom it comes, but also from where the recipient comes, where (s)he stands in that relationality. We should also try not to forget that even as the gift might remain an unknowable gift — *can we ever know what it means to give freedom to someone?* — it is passed: there is movement, *trans-*, transference, quite possibly a translation, transformation, of the data from one to another.

… knowledge economy and the imperative to produce knowledge of interest for the competitive war-games of the corporate world. As we know even academic fields where no patents can be produced have now been submitted to the general imperative of benchmark evaluation, have to accept the judgement of an academic market ruled by competition. We must admit that we have been asked to surrender a great part of our freedom to dissent; we have now to learn to our students to choose subjects leading to fast publication in high ranking specialized journals, about professionally recognized issues, that is in general, issues which interest nobody but other academic and fast publishing colleagues. One way or another we have to tell our students that if they want to survive, they have to format their questions, to translate them into academically acceptable, blindly normative frames …

… And we know that everywhere the same disempowering processes are at work. Everywhere a similar cut off from capacity to envisage, that is also, to feel, think and imagine, is produced. If there is a fight to be consented to, today, it might well be the fight for reclaiming this capacity, or even for reclaiming the capacity to envisage the possibility of reclaiming it.

And here, if one wanted an Agambenism, one could call it a movement from *zoë* to *bios*.

And the effect is a shift in her status—from a private entity, one within the realm of the home, *oikos*, to being a part of the public, a political entity, to her entering the *polis*.

But here, we should consider the possibility that there is a difference between being part of a *polis*, and becoming a part of a *community*: the former is a relationality of three, one involving the law, an external force on the relation; whilst the latter is that of two, a relation of exchange between those involved in the relation, shaping their own relation. Which does not mean that communities have no rules: they certainly do, and these might well be stricter than in a *polis*. For, one should bear in mind that one can negotiate, bend, play with, laws, due to the fact that they are overt, written, shared with all, governed, as it were, by a third, by something external to the parties involved, the people governed by the law; but with rules—especially since many remain subtle, hidden, covert—there is no space to read, negotiate, perhaps even no relationality without the rule itself; one either plays by them or is ejected. However, this does not erase the fact that it is the community itself that determines these rules—and even though one cannot deny that the laws of the land, the surrounding cultures, the *polis*, might have, probably has, effects on, even certainly does affect those rules, affects the community, if one is to attempting to think with precision, these differences should not be effaced. Whilst this is not the place to attend to that particular question, what is important to us here—what is relevant to the specificity of our space—is that community

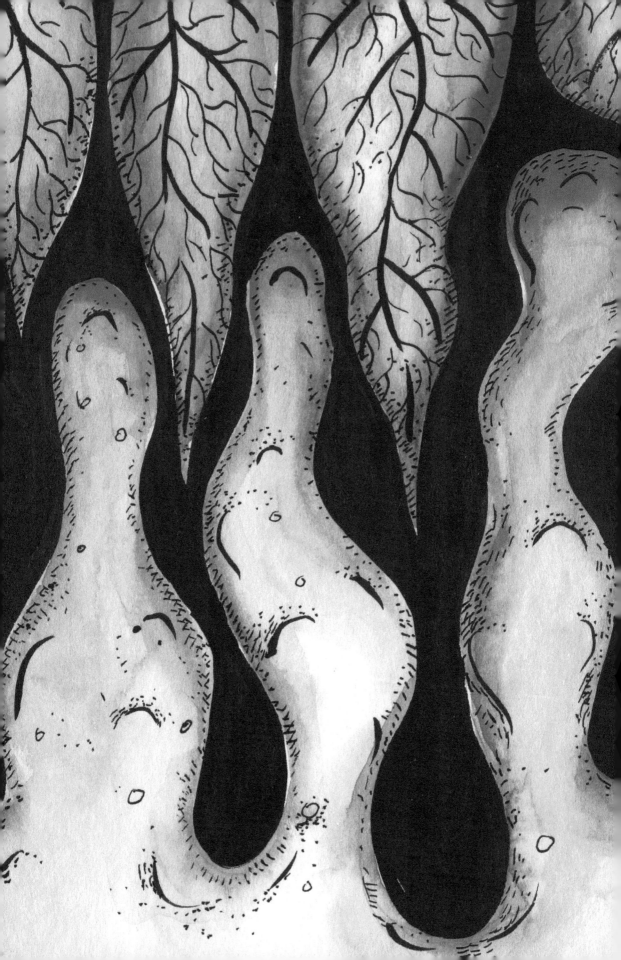

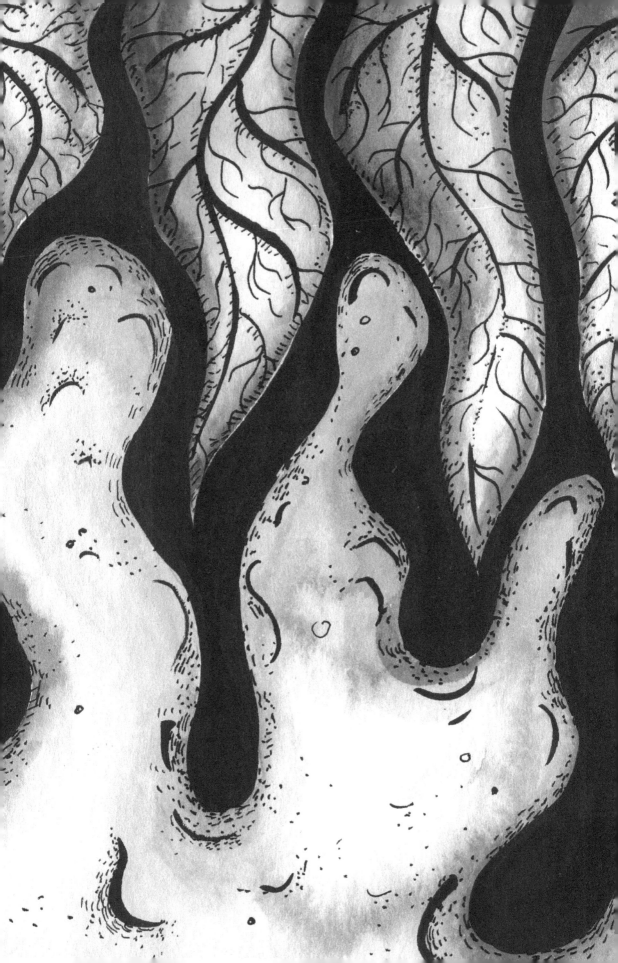

resounds with echoes of *munus*; also a gift, but an exchangeable gift, a gift of relation between two or more parties. In this exchange, a symbol is shared between the parties—a *symbolum* (creed, mark), is broken, after which all parties keep a part of it. This token (*symbolon*) becomes a binding agreement which carries through, moves through, generations, and is a watchword for the relationality between these parties; an agreement that brings the parties together (*sym-*) even as they move outwards (*bole*; throw, cast) into the world. Where the exchange—the actual object itself is not so important, but there has to be an object—of the gift itself is crucial. Which is not to say that this particular exchange is devoid of power: for, we should bear in mind that the *munus* is the foundation of potlatch: where generosity, more precisely performative generosity—ostentatious wastage really—functions as a means of putting others in their place.

However, this question remains: if a community requires *munus* and Science is, sciences are, so dependent on data, is the notion of a community of Science, a scientific community, possible?

To say *no* would be rather silly—of course it is.

But perhaps—and this is why Isabelle Stengers' address to scientists is important—this community, this communion, resides in the exchange between scientists, resides in scientists themselves, rather than Science itself.

Or, at least, that is its promise.

But, as we speak of promises, we should bear in mind Werner Hamacher's lesson from his beautiful

Whenever there is a promise, something other than the promise and something other than language—or simply another language—is also spoken. What is promised is always something other than understanding, other than another understanding, and other than an alteration of understanding alone. Something unpromisable.

Werner Hamacher, *Premises*

[8] Werner Hamacher. 'The Promise of Interpretation' in *Premises: Essays in Philosophy and Literature from Kant to Celan*, translated by Peter Fenves. Stanford: Meridian, 1999: 142.

[9] *Ibid*: 129.

essay, 'The Promise of Interpretation': that, in order to promise, there has to be something that is only to come, something not quite yet, something beyond; where the something that is promised cannot even have the status of a thing, or at least a known thing.[8]

Thus, there can never be a referent to the promise.

Which means that it is an utterance — since it has to be uttered or would never even exist — without any correspondence: **CATACHRESIS.**

Which does not only mean that it is potentially always already illegitimate, but that it might well occur outside of, external to, the utterance itself: completely other to the very utterance that attempts to give it the status of a promise.

As Werner continues, "interpretation is never the interpretation of a given other — whether it be a text, a person, a fact, an event, or a history — but it is always the laying out of what lays itself out in view of something else entirely."[9]

Where, a promise is precisely in its giving — a giving that quite possibility gives without an object, let alone an objective.

DATUM. But just because there is no necessary object does not mean that there is no recipient: without someone to receive, accept, acknowledge the giving, there is also no gift. Thus, even though both *datum* and *munus* entail power relations, not only is relationality its premise, the one receiving — the one at the short end of the stick, as it were — is the very condition of this play of power. And here, once again, it is not too difficult to hear echoes of daddy needing his followers in order to be kahuna.

After all — as Neil Gaiman has taught us — without devotees, even gods fade away.

But, since the relationality between the one giving and the one receiving is not quite so clear, or at least not so linear, this suggests that there might be a reversibility in the relation: that at the point of exchange—regardless of its status as *munus* or *datum*—the giver and the one receiving are momentarily exchangeable as well, or at least bound by the giving; quite possibly brought together (*sym*) by what is to be broken (*bole*). Thus, even as one might think that one is interpreting, the one interpreting is always already being interpreted—a being in interpretation only because of the possibility of interpretation.

And—tuning in once again to Werner—"if the one who activates interpretation is 'unhinged', so too is the interpretation as act. Interpretation is not a performative; it is not an act in the sense of a praxis performed by a subject, nor is it the deed of an empirical or transcendental doer, whether this doer is called will, grammar, or faith. Rather, interpretation is the aporetic—the self-missing—premise of every possible performance, a pre- and mis-performance. It has no Being, is not a transhistorical substance but a becoming without ground or goal, at the limit of 'itself', neither act nor fact, but—with all the unresolved tension this concept connotes—an *affect*. Interpretation is, in short, the word for the aporia of interpretation: for the experience of a nonsubjective process turning into a subject. And the experience is itself an aporetic one—hence an *affect*—because only at the point where a subject is not yet and will never be is it possible to undergo the experience of a still outstanding experience, the experience of an impossible experience and thus the experience of the impossibility of the aporia of experience."[10]

10 *Ibid*: 133.

[11] *Ibid*: 133.

[12] *Ibid*: 137.

Keeping in mind that the one who stands at the threshold is the stranger, the potential guest — this being the scene of, the very condition of the possibility of, hospitality itself.

Thus, interpretation is the promise of the possibility of interpretation. Keeping in mind that "promising means nothing else — a promise of the mere possibility of making promises."[11]

Thus, interpretation is premised on the promise; a promise that is hinged on the possibility of nothing but the possibility of interpretation.

Which is not to say that the one who is promising is not responsible for the promise. For, if it is premised on interpretation, the promise cannot exist without the subject — the I — who utters that promise, a promise that cannot be unless (s)he interprets it as such.

Here, we return to Hamacher, where he continues: "the possibility of infinite interpretations makes every particular interpretation and therefore the very concept of interpretation contingent. The possibility that the world, the perspective of the will, and interpretation could always be another one and, a *limine*, none at all — this potential of other possibilities that interpretation can never exhaust — inscribes an uncontrollable alterity into the very concept of interpretation and forbids, strictly speaking, all talk of interpretation *itself*. Tinged by other interpretations and non-interpretations, every interpretation must also be capable of being something other than interpretation and, a *limine*, no interpretation at all. Every interpretation is exposed to its other and to its Not: each one from the beginning an ex-posed, interrupted interpretation."[12]

Thus, in interpretation, in language, in thinking as such, we are always on a *limine*, threshold — where the very limit, boundary, is us. Not an us in the sense of a stable community, but in the very sense of a

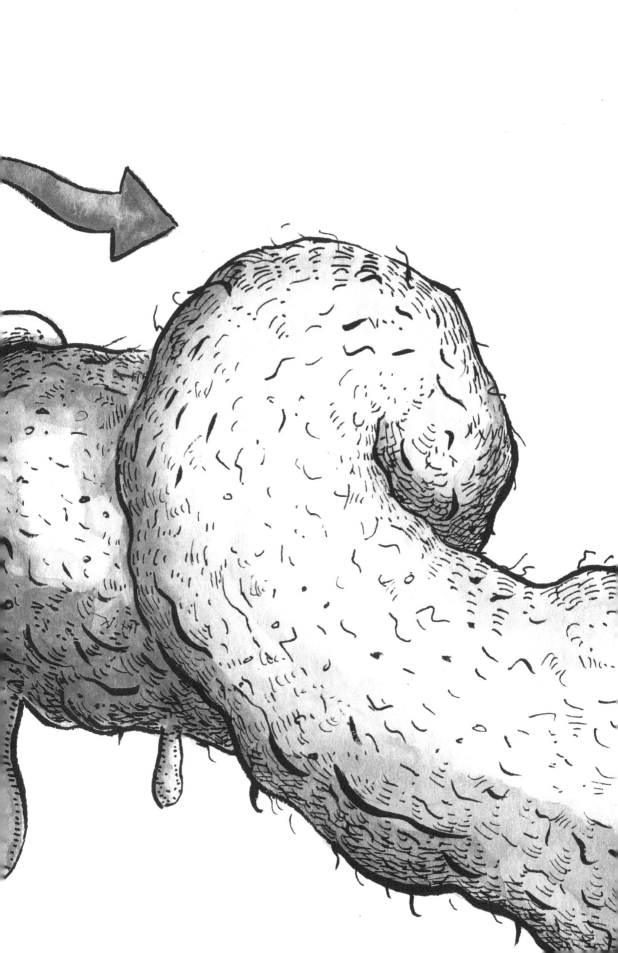

munus—through a symbolic agreement that binds, that throws us beyond ourselves, in a momentary interpretive promise, in a momentary promise of interpretation.

And if thrown, beyond us, outside us, then perhaps always also a *munus* haunted by the unknowability of a *datum*—a community that is interpreting itself as a community, even as it is attempting to read its communion into being. Thus, a community that is attempting to write itself—author itself—as a community, as it is reading itself.

And, since the community is composed of nothing more than the ones in the relationality with each other, this suggests that the site of the authoring is their very bodies. Hence, the authors and readers of the community are indistinguishable: and it is perhaps on this threshold that lies the very possibility of the community itself. And, it is in the space—the gap that allows the movement between writing and reading—where the coming together and breaking apart of the symbol can potentially take place.

But once we open the dossier that bodies are the sites on, perhaps in, which the communion is inscribed, we must once again reopen, readjust our receptors, to Nietzsche's warning. For, if written onto our *habitus*, there is always also the possibility that the community—no matter how unstable, continually changing it is—becomes a mere habit. And in this way becomes nothing more than *dogma*. Which is precisely what Stengers attempts to warn her audience of; the fact that thinking has been reduced to merely academically acceptable, blindly normative frames. Where what is lost—what we have been cut off from—is the possibility of taking ideas and their adventure seriously.

For, once this happens, what is destroyed is the very indiscernibility that is required for interpretation — the very basis of the possibility of the community itself. Thus, the moment the community of scientists knows — or thinks it knows — what Science is, the community itself dissolves. Thus, in order to maintain, think, itself as a community, the scientific community has to not only continually interpret, write, read, itself, it has to also be imagining, reimagining, what Science is. Which is not to say that nothing is done — of course much is, and with effects on the world, and the bodies of people in the world, but that each time something is done, each time there is the movement from imagining science to doing something with it, there is a leap of faith taken; not just in the outcome of the action, but in what science itself is.

Which also means that each time the community does something, enacts something in its name, it ruptures itself as community. For in re-imagining the Sciences, it is always also re-writing what the community of Sciences, of scientists — and hence — what community itself is. It is thus, a community that cannot avow to what it is to be a community — that only comes together, communes, in the very moment it writes itself, scribbles, *scribere*, scratches, tears, itself apart.

An unavowable community.

A community dreaming of the possibility of becoming a community.

Perhaps then, a dream community.

Which is why we should return and attend to Isabelle Stengers and her vigour in proclaiming:

I VITALLY NEED SUCH A DREAM, SUCH A STORY WHICH NEVER HAPPENED ...

And a dream in the sense of something unknown, something slightly beyond the boundaries, binds, of what is known. For, one must not forget that to begin to dream of something, it must first be within one's consciousness — if it were completely beyond one, one would not even be able to conceive of its dreamability. However, if it were completely within one's cognition, it is also no longer in the realm of a dream: it would merely be a prelude to an attempted actualisation (whether it can or cannot be actualised is perhaps irrelevant). Thus, for it to be dreamt it has to be both known and yet unknown, at exactly the same time. Which suggests that the story which never happened is not a tale outside, a tale completely exterior, alterior, to what has happened, is happening, but rather a never happened within what is happening, has happened. In other words, this story which never happened is what has been forgotten in memory itself; an unknowability within what is known.

Which is why Stengers' attempt to conceive of the community of scientists is not transcending the particularities of the so-called modern tradition, rather thinking with this particularity, rather trying to induce the capacity to imagine a possibility that it can be regenerated, or civilized. Which does not mean universalized. Rather, on the contrary, it means thinking with its own specific and dangerous, never innocent, ways of weaving relations, with the resources, imaginative, scientific, political, it may be able to activate in order to think with other Peoples and Natures. It is about attempting to respond with, whilst remembering that cosmopolitics has nothing

... just as I am sure that the reality of one night, let alone that of a whole lifetime, is not the whole truth ... no dream ... is just a dream ...

Arthur Schnitzler, *Traumnovelle*

to do with the miracle of decisions that 'puts everyone into agreement'. It rather concerns the demand that decisions be taken in full and vivid awareness of their consequences. Which is not to say that one can know in advance what the effects of one's decisions are: but that one accepts the responsibility for the decision, for the unknown in the decision, for what has been forgotten in them. For, this is the way in which decisions maintain an unknowability, maintain the question, within the momentary response — answer even — that decisions are.

Whilst trying never to forget that the question — along with the quest it opens, perhaps brings with it — is the very hallmark of Science itself.

And this brings us back to the very beginning, to the question that we — for, now that you are listening, reading, you are part of this particular communion — opened without actually opening: the question of *what, or even who, exactly is this Gaia*. After all, in order to compose with Gaia one has to at least momentarily have a notion of Gaia. But this is precisely what one has to resist, if one takes the notion of responding with Gaia seriously: for, the moment one has an idea of what or who she is, it is over. This is, of course, an impossibility in itself: for, if one has absolutely no idea what, who, one is responding to, with, one cannot then, cannot even, respond in the first place.

Which is perhaps why Tracy Bonham's echo of two mothers is crucial to us: if only to remind us that each time we respond to, with, *mother*, there is always already another mother; somewhere there, somewhere beyond — perhaps within — the first mother. A repetition that is the same, and that is somewhat *same same but different*.[13]

Which might be why there is no other way to end but to echo Stengers, read aloud Isabelle Stengers, whilst holding on to the possible impossibility that her words are also mine: where, what I have told you is just a tale, which as such, can certainly not hope to make a difference. But it calls for other tales…

Perhaps then, an ending that begs your indulgence — to restart, begin anew, read again, whilst continually attempting to attend to Isabelle Stengers' call, dream, call to dream. Which attempts to maintain the dream of this community of readers, listeners, who are attempting to listen to a dream.

A dream of thinking with: perhaps we could even call it a dream of momentarily being *in communion with Isabelle Stengers*.

A dream which dreams of the possibility of picking up Stengers' call — accepting her gift even when the fact of it being given remains unknown. Which attempts to steal away, perhaps even pilfer, her voice — and take flight. All whilst attempting to be faithful, attempting to listen to the cadence, sound, echo in — perhaps only through an act of attending, reading, rewriting; always already risking infidelity to — her voice. Whilst never quite knowing if it is even her voice — or just a sound:

GLAS

Or, as Tracy Bonham might say:

I'm losing my mind — everything's fine.

---

13 For a longer meditation on the phrase *same same but different*, please see my 'An Afterword — or, in the beginning there was …' in Jeremy Fernando, Jennifer Hope Davy, & Julia Hölzl. *[Given, If, Then]: a reading in three parts*. New York: Punctum Books, 2014, 30-31.

Resounding with echoes — with the spirit — of Jean Genet and his warning that, "all of this is only meaningful if I know that what I have just spoken is false."

... No, no,
the work of art
is not destined
for generations
of children.
It is offered to
the innumerable
population of
the dead.
Who accept it.
Or refuse it.
But the dead
of whom I was
speaking have never
been alive.
Or I forget the fact ...

Jean Genet

Contributors

Maureen Burdock is an award-winning feminist artist, graphic novelist and illustrator. She facilitates Laydeez do Comics San Francisco, a comics forum weighted towards women creators, which originated in the UK. Burdock has won several awards for her graphic novel work, including high commendation by the global Freedom to Create International Competition and top prize in the Judy Chicago/Through the Flower, Feminist Artists Under Forty Competition. The artist has received critical acclaim from diverse reviewers, including articles in *Marie Claire,* Mumbai, India; *Strip,* Copenhagen, Denmark; and the online publications *Lamp Project* and *Art Animal*. She has published reviews and articles in publications and catalogs such as *Graphic Novel Reporter, Art Practical,* and *WomanHouse v.4.0 Catalog*. Several gender studies and world literature professors have adopted Burdock's graphic novels for their classrooms, and McFarland published *Feminist Fables for the Twenty-First Century: The F Word Project* in 2015. Burdock continues to exhibit her work in gallery and museum venues, and the art for her current book, *Mumbi & the Long Run*, was exhibited at *Space Station 65 Gallery* in London in 2014. Her work is included in the Brooklyn Museum Elizabeth A. Sackler Center for Feminist Art Base.

Jeremy Fernando is the Jean Baudrillard Fellow at the European Graduate School, where he is also a Reader in Contemporary Literature & Thought. He works in the intersections of literature, philosophy, and the media; and has written twelve books — including *Reading Blindly*, *Living with Art*, and *Writing Death*. His work has been featured in magazines and journals such as *Berfrois*, *CTheory*, *TimeOut*, and *VICE*, amongst others; and he has been translated into Spanish and Slovenian. Exploring other media has led him to film, music, and art; and his work has been exhibited in Seoul, Vienna, Hong Kong, and Singapore. He curates the thematic magazine *One Imperative*; and is a Fellow of Tembusu College at the National University of Singapore.

Mariane Klettenhofer is an architect and graphic designer currently living in Brazil, where she works for the *Museum of Art of São Paulo Assis Chateaubriand — MASP*. She has previously worked mounting exhibitions for the photographer Sebastião Salgado; and with other artists, such as Artur Lescher — with whom she designed the book *Viajo porque preciso, volto porque te amo*, based on the movie of same title, directed by Karim Aïnouz and Marcelo Gomes. She has also designed books for children (*A Calorosa Aventura*) and the 3rd book of the collection *Arquiteturas*, dedicated to the architect Salvador Candia.

Lightning Source UK Ltd.
Milton Keynes UK
UKHW051259160922
408947UK00002B/127